NEW YORK CITY'S FINANCIAL CRISIS

Can the Trend Be Reversed?

Attiat F. Ott
Jang H. Yoo

American Enterprise Institute for Public Policy Research
Washington, D. C.

Attiat F. Ott is professor of economics at Clark University and an adjunct scholar of the American Enterprise Institute.

Jang H. Yoo is assistant professor of economics at Clark University.

336. 747
ᵩ 89 ₪

ISBN 0-8447-3195-1   *78-3542*

Domestic Affairs Study 40, November 1975

Library of Congress Catalog Card No. L.C. 75-39814

Printed in the United States of America

CONTENTS

NEW YORK CITY'S
FINANCIAL CRISIS

Introduction

Over the past several months, in daily news coverage, in the testimony of public officials and of business and financial analysts, the nation has been made aware of the financial crisis facing the nation's largest city. To anyone who has cared to read about New York, it has been made clear that New York City is on the brink of financial disaster and it has been argued that immediate federal action must be taken to prevent the financial collapse of the city if dire consequences are not to befall the nation and the world. Certainly the fact that New York City is facing a financial crisis is not to be disputed. What is in dispute is whether New York City's financial crisis is a local crisis that should be solved by local efforts, or a national crisis that warrants the attention of the President and the Congress.

How did the crisis come about? What are the warnings of imminent crisis that other cities similar in circumstances to New York City ought to watch for? Is the crisis temporary or of long duration? And what are the options? Those are the questions to which this study is directed. We begin with an overview of New York City's financial crisis. Next, we examine the roots of the economic crisis of our cities and of New York in particular. This examination is followed by a look at the fiscal behavior of seven major U.S. cities over the period from 1952 through 1973, in an attempt to determine whether they share similar or different experiences to that of New York. The analytical framework developed in this section is used

The authors wish to thank James Breen for running the 1975 AEI-BRAP model, Russell Cooper for his valuable assistance, especially in tracking down the necessary data for the paper, and Jared Lobdell for his editorial assistance.

1

to identify crisis signals in the cities examined. The next section of the study is devoted to the impact of New York City default. ("Default" here and throughout the study refers to a situation where either debt principal or interest is not paid off within thirty days after the due date.) Here we ask three questions: What would happen to the municipal bond market and municipal borrowing if New York City did in fact default? Would a default influence the liquidity position of major U.S. banks? What impact would a default have on the economy? On the basis of our findings, the fourth section of the study discusses some possible options for short- and long-term policy to deal with the financial problems of New York City. This section is followed by a brief conclusion.

1. New York City's Crisis: Some Current History

Although it is difficult to pinpoint the time when the financial problems facing New York City stopped being problems and started being a crisis, one may safely assume that the crisis had been reached early in 1975 when the city's bonds and notes could not be sold at any reasonable price. On March 13 and 20, 1975, the city, through its underwriters, offered for public sale $912 million of short-term notes at tax-exempt rates up to 8 percent. Yet, weeks after the offering, despite vigorous marketing efforts, the city could only sell $375 million of the notes, the rest remaining unsold. From that time on, the city was threatened with the possibility of default.

The Parameters of the Crisis. Why did the financial market suddenly close to New York offerings? To understand the market action, it is helpful to highlight some of the factors that led to the current crisis: (1) continuous budget deficits, (2) large and growing short-term debt, (3) cash flow problems, (4) the default of New York State's Urban Development Corporation, and (5) the credibility of the New York City government.

Budget deficit. In the period fiscal year 1967 through 1975, the accumulated budget deficits of New York City reached $2.6 billion with no prospects of closing the budget gap in sight. The revised budget proposed by Mayor Beame in his 1975–76 financial plan (see Table 1) put the fiscal year 1975–76 budget outlays at $12.3 billion, a reduction of $52 million over earlier estimates. Budget receipts were estimated to be around $11.7 billion, leaving a budget deficiency of $600 million.[1]

---

[1] Estimated by the Emergency Financial Control Board of New York State.

2

Table 1

NEW YORK CITY BUDGET, FISCAL YEAR 1975–76

($ billions)

Expenditures	
Welfare and charity	3.4
Education	2.5
Debt service	1.6
Health	1.1
Police	0.9
City university	0.5
Fire	0.4
Environment	0.3
Courts	0.2
Other	1.4
Total expenditure	12.3
Receipts	
General real estate tax collection	3.0
General fund revenues	4.0
Supplemental revenues—federal and state aid net of tax collected aid	4.4
Other revenues	0.3
Total receipts	11.7

Source: Expenditures from *Time* magazine, October 20, 1975, p. 15; receipts from *New York Times*, September 24, 1975.

Table 2 shows New York City budget deficits over the period from 1967 through 1975. As the table shows, the accumulated budget deficit has kept on growing. Furthermore, while the ratio of deficits to city revenues was 0.03 in 1967, it reached 0.07 in fiscal year 1973–74, making it increasingly difficult for New York City government to close the gap between expenditures and receipts.

Though the charter requires the city to write balanced expense budgets, successive mayors have done so only in a technical sense: throughout the period the city continued to cover the deficiency by short-term borrowing. Having borrowed to finance deficits and then lacking a surplus in later periods to pay off its borrowing, the only way New York City could pay off past loans was by floating new ones. As the deficits grew, the "borrowing" pyramid mounted—as of December 30, 1974, the outstanding short-term debt stood at $5.2 bil-

Table 2

**NEW YORK CITY RECEIPTS, EXPENDITURES AND
BUDGET DEFICITS,[a] FISCAL YEARS 1967–75**

($ millions)

	Receipts	Expenditures	Budget Deficit	Accumulated Budget Deficits
1966–67	4,346	4,461	−115	−115
1967–68	5,116	5,263	−147	−262
1968–69	5,968	6,044	−76	−338
1969–70	6,469	6,836	−367	−705
1970–71	7,072	7,359	−287	−992
1971–72	8,224	8,389	−165	−1,157
1972–73	9,166	9,686	−520	−1,677
1973–74	9,723	10,380	−657	−2,334
1974–75	11,911	12,201	−290	−2,624

a Total receipts include general fund, real estate taxes, grants from federal and state government, sales of property and other receipts (fees, charges, and similar receipts). Total expenditures cover operating expenditures, debt service, liquidation of encumbrances, and redemptions of city notes.
Source: Annual Reports of New York City Comptroller.

lion (about 18 percent of total municipal borrowing) and within six months, between December 1974 and June 1975, an extra $2.6 billion of short-term debt had been issued.

Short-term borrowing. Over the period from 1967 through 1974 New York City's short-term debt outstanding grew by more than 700 percent, or at a compound annual rate of slightly over 35 percent. This spectacular growth reflects the growth in two major components of the debt—tax anticipation notes (notes issued against taxes to be coming during the fiscal year) and revenue anticipation notes (notes issued against federal and state commitments for the fiscal year). The rate of growth of short-term debt outstanding and its composition is shown in Table 3.

In an examination of the components of New York City short-term debt, it is worth noting that the amounts of tax anticipation notes and revenue anticipation notes have been rising at extraordinary rates since 1967. Between 1967 and 1974, tax anticipation notes outstanding grew by about seven times their value in 1967, a compound annual rate of increase of 35 percent. The growth of revenue

4

Table 3

OUTSTANDING NEW YORK CITY SHORT-TERM DEBT, AT YEAR-END, 1967, 1974, AND FY 1975

($ millions)

	1967	1974	Annual Compound Growth Rate, 1967–74	1975
Tax anticipation notes	$136.5	$1,102.0	+34.8%	$ 380.0
Revenue anticipation notes	93.8	3,050.0	+64.4%	2,560.0
Urban renewal notes	29.9	107.6	+20.1%	30.0
Bond anticipation notes	374.7	1,008.7	+20.4%	1,570.2
Total short-term debt	$634.9	$5,268.3	+35.3%	$4,540.2
Ratio of revenue anticipation notes/revenue	0.042ᵃ	0.260	+35.5%	—
Ratio of tax anticipation notes/tax revenues	0.054	0.053	no change	—

ᵃ For 1968.

Sources: For 1967 and 1974 data, U.S. Congress, Senate, Committee on Government Operations, *Intergovernmental Anti-Recession Assistance Act of 1975*, Hearings before the Subcommittee on Intergovernmental Relations, 94th Congress, 1st session (May 6-8 and June 3, 1975), p. 174; for 1975 data, City of New York, *Three-Year Financial Plan for 1975–1976 through 1977–1978*, Financial Data, Schedule E.

anticipation notes is even more spectacular. Between 1967 and 1974 they increased by over 3,000 percent or at an annual compound rate of 65 percent. The growth of bond anticipation notes to finance construction (the notes being ultimately converted into permanent debt) was less spectacular—about 21 percent compounded annually during the period from 1967 through 1974.

To put these figures into proper perspective, we may compare the growth of the short-term debt to the growth of the revenues against which the city was allowed to borrow. As the table shows, the ratio of tax anticipation notes to taxes is about the same in 1967 and 1974, but the ratio of revenue anticipation to revenues rose from 0.042 in 1968 to 0.260 in 1974. It must, however, be noted that the two major components of New York City short-term debt—tax and revenue anticipation notes—have declined from their peak value of $4.0 billion (as of December 1974) since a part of these obligations matured at the end of the 1975 fiscal year. After repayment of

5

maturing short-term debts and the addition of new borrowing between December 1974 and June 30, 1975, short-term debt as of July 1, 1975 stood at $4.5 billion, of which $2.9 billion was in the form of tax and revenue anticipation notes. The three-year financial plan for fiscal year 1975–76 through fiscal year 1977–78 put the interest cost on temporary debt at $252 million for the fiscal 1975–76 budget, projected to decline to $153 million by fiscal 1977–78. Annual interest costs increased from $177.9 million in 1967 to $463.5 million in 1974.

Although bond anticipation notes (BANs) are of less immediate concern than temporary debt, they too have contributed to the current crisis. Instead of these bonds being funded by long-term borrowing, it was decided (because long-term interest rates were high in comparison to short-term interest rates) that the city would borrow short-term and that each year the debt would be "rolled over." Unfortunately the expected savings from this course of action did not materialize. The last issue of BANs cost almost 8.7 percent and put the city in a position where it would have to raise taxes in 1975–76 to cover the difference between the 8.7 percent cost and the anticipated revenues, or go once again to the money market to cover the deficiency.[2] For fiscal year 1975–76, interest on anticipation notes is estimated to be about $127 million or 3.1 percent of the value of notes outstanding. Table 4 shows outstanding short-term debt notes as of July 1, 1975 and debt service as of September 1, 1975.

Cash-flow problems. Although the city was running at a deficit over the past several years, the possibilities of default became apparent when the city found itself illiquid early in 1975. The cash-flow problem came to a head as the result of a law suit questioning the issuance of Stabilization Reserve Corporation bonds in February 1975. Faced with a deficiency in cash for paying off $308 million on budget notes that could no longer be rolled over, the city obtained legislation to create the Stabilization Reserve Corporation (SRC) with power to borrow $520 million to be turned over to the city to liquidate its notes. In return the city would be required to pay annually to the SRC, in December, a sum equal to annual interest and amortization costs. In the event that the city was unable to make the December payment, the state comptroller was to pay the sums out of state-collected city revenues or directly from the state's own funds.

2 U.S. Congress, Senate, Committee on Government Operations, *Intergovernmental Anti-Recession Assistance Act of 1975*, Hearings before the Subcommittee on Intergovernmental Relations, 94th Congress, 1st session (May 6-8 and June 3, 1975), p. 176.

Table 4

NEW YORK CITY SHORT-TERM DEBT:
INTEREST AND PRINCIPAL

($ millions)

	Interest	Principal
Short-term debt Outstanding (7/1/75):		
Bond anticipation notes	$126.9 a	$1,570.2
Other	251.9 a	2,970.0
Total	$378.8 a	$4,540.2
Short-term debt Due before 12/31/75:		
Due November 10 b	19.3	250.0
Due December 11 b	30.7	400.0
Due December 17 c	2.3	30.0
Total	$ 52.3	$ 680.0

a As of September 1, 1975.
b Revenue anticipation notes.
c Urban renewal notes.
Source: For debt outstanding and for principal of debt maturing, City of New York, *Three-Year Financial Plan for 1975–1976 through 1977–1978*, Financial Data, Schedules E and F; for interest on debt maturing, authors' estimates.

The timing of the suit could not have been much worse for New York City, inasmuch as when it was filed the city had already paid off the budget notes and the cash-flow problem was intensified. Unable to issue SRC bonds or find alternative resources, the city plunged deeply into the use of capital funds. During the fiscal years 1972–73 and 1974–75, the city spent over $1.5 billion of capital funds to cover operating expenses. For fiscal 1975–76 the capital budget is expected to be cut back to $1.1 billion.[3] Between December 1, 1975 and June 30, 1976 the city's cash needs are estimated to be about $4 billion, of which $500 million is needed for meeting current expenses.

Insolvency of the Urban Development Corporation. The brief slip of the New York State Urban Development Corporation into insolvency, in February 1975, played a major part in the decision of the banking community to reject New York City's bids for new short-term loans. New York banks that in the past had muted their com-

[3] *U.S. News & World Report*, September 15, 1975, p. 30.

plaints about New York City fiscal "gimmickry" while collecting high interest rates on city debt began flashing a stop sign. In March, with the financial markets closed to New York City, the municipal government was kept afloat only by prepayments from New York State of some $800 million in aid due July 1, 1975. Furthermore, in June, the state stepped in with the creation of the Municipal Assistance Corporation—known as Big Mac—to serve as an interim borrowing vehicle for the city. Big Mac was authorized to issue up to $3 billion in notes for the city's benefit, thus covering the city's cash deficits for the July–September period. The Municipal Assistance Corporation was set up as an agency for New York State; its bonds have state backing and the city's debt service costs are to be met from sales and stock-transfer taxes. In spite of these assurances, the Municipal Assistance Corporation encountered difficulty in borrowing for the city. The corporation raised only a patchwork $2 billion and then ran into investor resistance. In August, Big Mac was able to borrow less than half of its planned amount, even though the new issues carried interest rates up to 11 percent.

The credibility of New York City government. Adding to New York City's financial problems was the attitude exhibited by the city government towards its financial crisis. Early in the game, the mayor declined to face the possibility of default, insisting that the city was able to meet all its obligations. When questioned on fiscal year 1975–76 budget deficits and the city's plans to meet the deficiency, the mayor failed to give city creditors what they considered a viable solution to the problem, as well as failing to show a willingness on the city's part to exercise the budget discipline needed to meet the crisis. As early as December 1974, the mayor had called for layoffs and service cuts to reduce the size of the 1974–75 deficits, but most of the resolve went into the press release.[4] Investors heard the mayor call for 50,000 layoffs, then read that only a third of those layoffs had been carried out. As the mayor himself put it, "the big credibility crisis really was the one where I submitted the budget and said 51,000 employees are going to be dropped—knowing full well that we're not going to drop 51,000!"[5]

The uncertainty of city action did not diminish as the crisis took shape in March. During the six months from March to September hard figures on the city's debt, cash reserve, payrolls, number of workers laid off, and other significant variables, fluctuated with

---

[4] *Wall Street Journal,* September 4, 1975, p. 10.

[5] *New York Times,* Sunday, October 5, 1975.

"mind-boggling regularity."[6] The mayor's plan for budget cutting changed from day to day. In fact the mayor did not announce the actual budget trimming until he was forced to do so by the Emergency Financial Control Board.[7] The mayor's new estimate of budget cuts for fiscal 1975–76 is about $52 million with a layoff (through attrition) of 6,248 full-time workers.[8] His announced budget deficit for fiscal year 1975–76 is about $724 million which is slightly smaller than his earlier forecasts of an $800 million deficit, but larger than the $600-$700 million talked about in early October.[9] In addition to the frequent changes in budget numbers, the attitude of city government toward the state rescue plan was in itself damaging to investor confidence. The well-publicized squabbles among city officials, union leaders, the banks, and members of the Municipal Assistance Corporation contributed in no small way to damaging the credit of the corporation and the credibility of the city itself. As one Big Mac official put it, "The City has had to be dragged, kicking and screaming, to do what had to be done."[10]

The behavior of the mayor can also be faulted. Not only did he fail to call for fiscal restraint, not only did he fail to be specific on budget cuts, but he openly resisted the Emergency Financial Control Board's moves to deal with budget deficits. In a letter to the governor, Mayor Beame wrote that the "financial plan" would provide New York City with a balanced budget at the end of the 1977–78 fiscal year, but that the price in terms of the impact on the city was "exorbitant."[11]

Is Default Imminent? Deadlines for New York City default seem to have loomed up and then disappeared overnight as some new saving grace was discovered or manufactured. A major crisis came on October 17, and the city was able to avert default in the eleventh hour as the United Federation of Teachers agreed to invest $150 million of its pension funds in Municipal Assistance Corporation bonds. Although the immediate crisis was averted, the question of default remains. Between December 1 and December 11, 1975, the city must come up with $644.5 million to meet its payroll and debt service

---

[6] *Wall Street Journal*, September 4, 1975, p. 10.

[7] The Emergency Financial Control Board was created by the state on September 9, 1975, and charged with administering the city's finances.

[8] City of New York, *Three-Year Financial Plan for 1975-1976 through 1977-1978.*

[9] *New York Times*, October 5, 1975.

[10] *Time* magazine, October 20, 1975, p. 11.

[11] *Wall Street Journal*, October 16, 1975, p. 6.

obligations. Thus we pose the question: can the city permanently avert default or is default imminent?

On this issue there are widely divergent opinions. There are those like Congressman Benjamin Rosenthal who believe that "without federal assistance the largest bankruptcy in U.S. history is imminent," [12] while others, notably Secretary of the Treasury William Simon, discount the possibility of default and are optimistic about New York State efforts (and the efforts of Big Mac) in averting default. To ascertain whether default can be averted the next time around, it is useful to look at the potential sources of funds on which the city can draw. There are five possibilities, each of which can help New York City escape default: (1) a federal guarantee of New York City loans, (2) congressional action on some form of federal financial assistance, (3) investment of New York City pension funds in Big Mac bonds, (4) investment of New York State pension funds in Big Mac bonds, and (5) additional borrowings from the financial market through Big Mac or through the forced restructuring of outstanding debt.

The city has already had some success in tapping the last three of these sources. On August 21, 1975, the city, although with some difficulty, succeeded in borrowing $792 million from the financial market to meet its debt payments due the next day. On October 17, 1975, the city, again on the edge of default, was able to borrow $150 million from its teachers' retirement system. This borrowing, together with borrowings from state and other pension funds, enabled the city to meet $478 million of immediately maturing debts and interest.[13] For the remainder of the year, the city will need $732.3 million for maturing debt services, as outlined in Table 4.

In effect, of $1.5 billion of total debt service for 1975, the city had met only one-half as of October 31. During November, cash needed to retire debt will presumably be met by a Big Mac state aid package, leaving about $430.7 million for the city to raise by December 11, 1975. To finance the remaining part of the maturing debts through December, its future debt service, and its budget expenses

---

[12] Opening statement of Representative Benjamin Rosenthal (Democrat, New York) during the "Oversight Hearings into Adequacy of Federal Agency Studies of National Impact of New York City Default," before the Subcommittee on Commerce, Consumer, and Monetary Affairs which he chairs, October 8, 1975.

[13] The aid package consisted of $250 million from the state pension system, $33 million from the state's sinking funds, $100 million from the city's other pension system, and $95 million from the teachers' retirement system. Some $55 million more was provided by the teachers' retirement system for debts maturing on November 1, 1975.

Table 5

STATEMENT OF NEW YORK CITY CASH NEEDS, DECEMBER 1, 1975 THROUGH JUNE 30, 1976
($ millions)

Month	Expense Budget and Capital Cash Flow	Long-term Principal	Long-term Interest	Short-term Principal	Short-term Interest	Total
Dec.	$489	$ 7	$ 9	$ 430	$ 40	$ 975
Jan.	314	117	49	820	77	1,377
Feb.	84	180	45	290	22	621
March	404	61	30	491	43	1,029
April	(11)	123	42	—	—	154
May	(274)	67	28	220	16	57
June	(525)	1	8	332	26	(158)
Total	$481	$556	$211	$2,583	$224	$4,055

Source: City of New York, *Three-Year Financial Plan for 1975–1976 through 1977–1978*, Financial Data, Schedules E and F.

for the remainder of fiscal year 1975–76 (through June 30, 1976), some $4 billion will be needed. Table 5 shows New York City's cash needs for December 1, 1975, through June 30, 1976.

Looking at the possibilities open to the city, one must be skeptical of expecting a federal bailout. The Ford administration has repeatedly refused to extend aid to New York City. The President has made it clear that he remains opposed to federal rescue.[14] It is reasonable to assume that, at least in the short term, a sudden shift in his administration's attitude towards federal aid to New York City is unlikely to take place. This leaves open the possibility of congressional action before the next deadline for default in December 1975. However, it appears unlikely (as of early November) that Congress will act quickly on an aid formula. Even if congressional aid were to come, it would probably come too late for the December deadline. Of course, it is likely that if a federal bailout were to come, the price attached would be a requirement that the city put its house in order. All the proposals made to the various congressional committees call on the governor of the state of New York to give

---

14 *Wall Street Journal*, October 24, 1975.

Table 6
BREAKDOWN OF NEW YORK STATE PENSION FUNDS
HOLDINGS, BY TYPE OF ASSET
(as of December 31, 1975)

New York State Common Retirement Fund

U.S. government bonds	$ 420,400,000
U.S. government agencies	316,500,000
State and municipal bonds	38,900,000
Canadian obligations	114,500,000
Corporate bonds and railroad equipment trust certificates	1,386,200,000
International Bank (World Bank), Inter-American Development Bank, and Import Bank	57,800,000
Mortgages	1,019,500,000
Common stock, book value (market value $983,400,000)	830,400,000
Miscellaneous short-term investments	275,100,000
Total assets	$4,459,300,000

New York State Teachers' Retirement System

Bonds:	
U.S. government	$ 27,238,100
States, territories and possessions	50,000
Canadian	127,274,027
Political subdivisions of states, territories and possessions	18,188
Special revenue and special assessment obligations and all nonguaranteed obligations of agencies and authorities of governments and their political subdivisions	
U.S. government	75,887,315
Canadian	10,413,500
Railroads	51,594,106
Public utilities	900,539,355
Industrial and miscellaneous	535,409,752
Stocks:	
Railroad	$ ————
Public utilities	78,812,020
Banks, trust and insurance companies	10,203,368
Industrial and miscellaneous	468,983,189
Total bonds and stocks	$2,286,422,920
Real estate holdings:	
Conventional mortgages	598,757,177
FHA mortgages	107,379,733
Total assets	$2,997,468,272

Source: Richard Tucker, *State and Local Pension Funds 1972* (New York: Securities Industry Association, 1973).

assurances that the city would "balance" its budget by 1978 and (perhaps) restructure its outstanding debt.[15]

If we rule out the possibility of federal assistance and if we assume the municipal market will remain temporarily closed to New York City's direct borrowing, the options remaining are for the city to persuade its own pension funds and the state's pension funds to invest in city notes, and to turn to Big Mac for additional borrowing on its behalf. As of December 1972, cash and security holdings of New York state and local government retirement systems amounted to $14.9 billion.[16] As of December 31, 1971 the total assets of two New York state pension funds—New York State Common Retirement Fund and New York State Teachers' Retirement System—were $7.5 billion. A breakdown of the two pension funds' assets by type of holding is given in Table 6. In the three-year financial plan submitted by the city to the emergency board the city proposed to solve its cash needs by a combination of Municipal Assistance Corporation bonds, the use of pension funds, aid advances and city capital debt issue, as shown in Table 7.

Whether New York City is heading towards default this December or some time in the near future clearly depends on the city's ability to tap retirement funds and restore market confidence in the worthiness of Municipal Assistance Corporation notes. In the long run, however, the financial stability of the city hinges on its ability to balance its budget, convert a major portion of its short-term debt into long-term debt and expand its tax base in proportion to its services (or shrink its services in proportion to its tax base). Failure of the city to meet its cash needs through fiscal 1975–76 would surely culminate in a default.[17]

The consequences of a default by the city of New York—between December and June—and its repercussions on the financial market and the economy, will be discussed in a later section of this paper.

---

[15] See section on Policy Options, below.

[16] David J. Ott, Attiat F. Ott, James A. Maxwell and J. Richard Aronson, *State-Local Finances in the Last Half of the 1970s* (Washington, D. C.: American Enterprise Institute, April 1975), p. 88.

[17] It is hard to foretell, but since the state's emergency control board (consisting of the governor, mayor, state comptroller, city comptroller, and three gubernatorial appointees) holds the right to reject a considerable part of union contracts which include various contingent factors, it may be possible for the city to gain the use of the funds through persevering negotiations. If this is the case, then default may not be imminent.

Table 7

NEW YORK CITY FINANCIAL PLAN TO MEET CASH NEEDS FROM DECEMBER 1, 1975 THROUGH JUNE 30, 1976

($ millions)

	MAC a	Rollovers— Banks, Pension Funds, Sinking Funds	Sale of Mitchell- Lama Mortgages	Short-Term Debt to Be Redeemed	Aid Advances and/or City Capital Debt Issues	Total
December	—	$ 60	$300	$609	$ 6	$ 975
January	$ 500	210			667	1,377
February	250	30	150		191	621
March	250	48	150		581	1,029
April					154	154
May					57	57
June		451		(609)	—	(158)
Total	$1,000	$799	$600	$ 0	$1,656	$4,055

a Additional MAC issues produce a total $4.1 billion MAC issuance. Of this total, $3.6 billion will finance prior and 1975–76 deficits. The remaining $500 million will finance city capital needs. It is therefore assumed that the Municipal Assistance Corporation will purchase $500 million of city bonds.
Source: See Table 4.

2. The Roots of the Crisis in Urban Finance

It is generally agreed that a variety of factors are responsible for the fiscal crisis facing New York City today. Some of these can be classified as external to New York City while others are internal and of its own creation. External factors are those associated with the national economic health. Over the past two years, the U.S. economy has experienced the ills of inflation and recession together. The inflation-recession combination that was in large part responsible for the sharp rise in the cost of city services and welfare expenditures contributed at the same time to the deterioration in city revenues. In this regard, New York City was perhaps more vulnerable than most municipalities because New York (like other large cities) relies more heavily on sales and income taxes than do most units of local government, whose revenues come primarily from property taxes. Sales and income tax bases tend to be more susceptible to cyclical variations than the property tax base. As Table 8 shows, income and sales taxes account for 24 percent of New York City's total tax receipts in 1974.

Table 8

NEW YORK CITY INCOME, SALES, AND REAL ESTATE TAXES AND YEAR-TO-YEAR
CHANGE, FISCAL YEARS 1969–74

Period	Income and Earnings		Corporate and Business		Sales		Real Estate		Economic Condition
	$ Millions	Percent change	$ Millions	Percent change	$ Millions	Percent change	$ Millions	Percent change	Index
1969–70	205.5		245.5		466.9		1,893		101.3
1970–71	199.4	−3.0	221.6	−9.7	493.6	+5.7	2,080	+9.9	110.1
1971–72	443.2	+122.3	375.1	+69.3	519.7	+5.3	2,189	+5.1	125.0
1972–73	439.6	−0.8	368.0	−1.9	551.3	+6.1	2,468	+12.7	119.8
1973–74	454.8	+3.5	363.6	−1.2	575.4	+4.4	2,655	+7.6	94.8

Source: Tax data from the *Annual Report* of the Comptroller of the City of New York, for the fiscal years 1969–70 to 1973–74. The economic condition index (composite of twelve leading indicators) from U.S. Department of Commerce, Social and Economic Statistics Administration, Bureau of Economic Analysis, May 29, 1975.

However, current economic conditions cannot be entirely blamed for the New York City crisis. The roots of the New York City crisis go far deeper than any transitory national economic problems. New York, like most urban centers, has been affected by increased pressure for more public services, while at the same time the tax base has been eroded and with it the city's capacity to meet the ever-rising demand for services. The following common factors are said to have caused the fiscal ill health of the major U.S. cities: erosion of the tax base, influx of unskilled poor and minority groups, decay and obsolescence of the center city, unionization of public employees, growing demand for anti-congestion and pollution expenditures, and the short-run horizon (myopia) of public officials.

The analysis of cross-section data for most large U.S. cities tends to supply documentation for most of these causes. In their 1973 report on city financial emergencies, the Advisory Commission on Intergovernmental Relations (ACIR) pointed out:

> Large cities have been operating in an increasingly difficult environment. The sources of stress often include declining population, the impact of collective bargaining by city employees, the impact of inflation on the costs of many labor-intensive services that city residents desire, a lack of growth in the tax bases of many central cities, and the slow growth in many of the cities' traditional sources of revenue. These and similar factors combine to create a fiscal and political tightness in the financial affairs of cities that makes them increasingly susceptible to financial emergencies.[18]

These problems are compounded for New York City because of its location (port of entry to many immigrants), size and density, and because of its "liberal" attitude towards public services. To put the underlying causes in proper perspective we first present, in Table 9, the main features common to most urban cities. Next, in Table 10, we concentrate on some of those conditions not generally experienced by other cities—conditions which could explain why New York City and not, for example, Chicago or Houston is facing a crisis today.

The Characteristics of Seven Major Cities. From Table 9, it is clear that the New York experience is not unlike that of the other cities being considered. With the exception of expenditures on public welfare, where only New York City and Denver bear such a burden, all seven cities exhibit an increasing ratio of minority population,

---

[18] Advisory Commission on Intergovernmental Relations, *City Financial Emergencies: The Intergovernmental Discussion* (Washington, D. C.: Advisory Commission on Intergovernmental Relations, July 1973), p. 56.

16

Table 9

DEMOGRAPHIC AND OTHER CHARACTERISTICS OF NEW YORK CITY AND SIX OTHER U.S. CITIES, 1970–73

City	Year	(1) Population Total (thousands)	(1) Population Percent non-white	(2) Income Per capita	(2) Income Central city/ outside central	(3) Value of Houses: Central City/ Outside Central	(4) Public Employment per 1,000 Population	(5) Percent of Total Employment	(6) Welfare Payment per Capita	(7) Average Salary of Public Employee (monthly)
New York	1970	7,894	13	$4,038	.88	.82	47.5	11.7	$206.2	$ 870
	1973	7,775	20	4,815	.94		52.5		314.0	1,008
Chicago	1970	3,356	23	3,804	.85	.82	13.1	3.2	3.5	903
	1973	3,291	33	4,590	.89		32.7		3.3	959
Detroit	1970	1,519	29	3,639	.94	.71	17.0	4.6	3.3	801
	1973	1,462	44	4,511	.87		30.7		2.7	1,060
Cleveland	1970	742	29	3,097	.70	.68	21.0	5.5	0	622
	1973	683	38	3,726	.77		34.9			792
Buffalo	1970	456	5	3,303	1.22	.68	29.3	7.8	0	812
	1973	441	8	3,818	.88		36.5			887
Denver	1970	527	6	3,654	1.10	.88	16.8	4.2	56.4	709
	1973	514	9	4,812	1.02		29.8		94.7	758
Houston	1970	1,325	23	3,559	1.15	.99	7.5	1.8	0	403
	1973	1,325		4,378	1.03		28.0			679

Note: Because the 1970 data in column (4) are from one source and for the central city areas and the 1973 data from another source and for the standard metropolitan statistical areas, no comparisons may be made between the two years, though comparisons may be made among cities in each year.

Sources: Columns (1) and (2) from Sales Management, Inc., *Survey of Buying Power* (1971, 1974); columns (3) and (5) from Advisory Commission on Intergovernmental Relations, *City Financial Emergencies: The Intergovernmental Discussion* (Washington, D. C.: ACIR, July 1973); column (4) from Department of Commerce, *City and County Data Book 1972* (Washington, D. C., 1973) and *Local Government in Metropolitan Areas* (same) for 1970, and from Sales Management, *Survey of Buying Power* (1974) for 1973 (see note above), data for standard metropolitan statistical areas; columns (6) and (7) are from Department of Commerce, *City Government Finances in 1969–70* and *1972–73* (Washington, D. C., 1971 and 1974).

Table 10

COMPARISON OF FISCAL PERFORMANCE OF NEW YORK CITY AND SIX OTHER U.S. CITIES, 1970–73

DOLLARS

	New York		Chicago		Detroit		Cleveland		Buffalo		Denver		Houston	
	1970	1973	1970	1973	1970	1973	1970	1973	1970	1973	1970	1973	1970	1973
Per capita total expenditure	1,002	1,483	235	313	315	466	280	349	386	595	336	535	135	184
Per pupil education expenditure	906	1,839	0	0	0	0	0	0	n.a.	398	0	0	0	0
Per capita other expenditure	786	1,183	235	306	312	459	289	349	226	341	336	534	135	181
Per capita total revenue	970	1,452	218	336	335	536	277	379	365	591	341	527	131	192
Per capita tax revenue	383	529	113	169	147	193	129	134	145	206	133	199	81	104
Per capita federal-state aid	383	626	42	70	65	183	23	82	139	264	95	159	3	22
Per capita short-term operating debt	183	355	78	79	56	65	83	111	175	132	28	35	0	0
Per capita general obligations bonded debt	828	1,612	86	88	171	202	171	212	229	370	295	426	257	280
Per capita assets	1,061	1,275	239	346	424	615	173	239	220	260	244	410	180	215
Budget surplus or deficit (−), in millions	−254	−248	−59	73	30	103	−9	21	−10	−2	2	−4	−4	11

Table 10 *(continued)*

PERCENT

	New York 1970	New York 1973	Chicago 1970	Chicago 1973	Detroit 1970	Detroit 1973	Cleveland 1970	Cleveland 1973	Buffalo 1970	Buffalo 1973	Denver 1970	Denver 1973	Houston 1970	Houston 1973
Budget surplus/Total expenditure	−3.2	−2.1	−7.5	7.1	6.2	15.1	−3.2	8.7	−5.5	−0.5	1.3	−1.5	−2.4	4.3
Budget surplus/Total revenue	−3.3	−2.2	−8.1	6.6	5.9	13.2	−4.4	8.0	−5.8	−0.5	1.3	−1.5	−2.5	4.1
Tax revenue/Total expenditure	38.2	35.7	48.1	54.0	46.7	41.4	46.1	38.4	37.6	34.6	39.6	37.2	60.0	56.5
Federal-state aid/Total revenue	39.5	43.1	19.3	20.8	19.4	34.1	8.3	21.6	38.1	44.7	27.9	29.8	2.3	11.5
Federal-state aid/Total expenditure	38.2	42.2	17.9	22.4	20.6	39.3	8.2	23.5	36.0	44.4	28.3	29.7	2.2	12.0
Short-term debt/Total revenue	18.9	24.4	35.8	23.5	16.7	12.1	30.0	29.3	47.9	22.3	8.2	6.6	0.0	0.0
Short-term debt/Total assets	17.2	27.8	32.6	22.8	13.2	10.6	48.0	46.4	79.5	50.8	11.5	8.5	0.0	0.0
General obligation bonded debt/Total assets	78.0	126.4	36.0	25.4	40.3	32.8	98.8	88.7	104.1	142.3	120.9	103.9	142.8	130.2

Note: See qualification on p. 16. Figures are not strictly comparable because New York, as a city comprising five counties, is liable for welfare and education expenditures for which the others (except Denver) are not liable. The figures may however be compared if this qualification is kept in mind.

Source: U.S. Bureau of the Census, *City Government Finances in 1969–70* and *1972–73* (Washington, D. C.: Government Printing Office, 1971 and 1974).

19

a declining population (with the exception of Houston), a ratio of per capita income in the central area to per capita income outside the central area (except for Houston and Denver) of less than one (declining for Buffalo and Detroit), and increases in the number of public employees per 1,000 population.

To compare the fiscal performance of New York City to that of the other six cities chosen, we show in the first part of Table 10 per capita expenditures, tax revenues, federal and state aid and short-term operating debt for the seven cities for 1970 and 1973. As the table shows, the level of New York City's expenditures, taxes and borrowing is far above the level for any of the other six cities examined. Furthermore, in 1970 and 1973, New York City received a higher share of state and federal aid than any other city in the sample.

In the second part of Table 10, expenditures, taxes, aid, and debt are related to financial and tax capacity. Looking at the ratios of tax revenue to expenditures, short-term debt to total revenues and total assets, we find that although all the cities examined exhibit similar "environments" they have differed considerably in their financial performance. New York City, with the highest per capita expenditures and taxes, has (along with Buffalo) the lowest tax effort ratio: its ratio of tax revenues to total expenditures was 38 percent in 1970 declining to 35 percent in 1973, while Houston had over a 50 percent tax effort ratio for both years. Short-term debt as a percent of total revenues and total city assets for New York City, though not so high as for Buffalo, had increased between 1970 and 1973, while these two ratios had declined for all other cities. Federal and state aid is found to favor New York City. In 1973 the ratio of federal and state aid to total New York City expenditures was 42 percent. (Remember that the ratio of city taxes to expenditures was also only 35 percent in 1973.) Other cities in the sample (again with the exception of Buffalo) have a higher ratio of taxes to expenditures than New York City has and a lower ratio of federal and state aid to expenditures, because most of these cities do not provide welfare services, and many have separate educational funding.

Chicago and Houston, with much higher concentrations of non-whites than in New York City, show a far better fiscal performance than New York City. The tax effort ratios in the two cities are around 1.5 times the ratio in New York City. In addition, Houston had no short-term debt in either 1970 or 1973 and the reliance of both cities on federal and state aid to cover expenditures is only about one-third to one-half that of New York.

The poor fiscal performance of New York City is again illustrated in Table 11. Using six representative indicators, we find that New York City's overall budget management puts it last among the seven largest cities considered. On the revenue side, the city's per capita income grew at the annual rate of only 6 percent from 1970 through 1973, while per capita expenditures grew at 14 percent. The city's tax effort is low compared to that of other cities. The budget deficit has decreased (or the budget surplus increased) in many other cities, but New York City's deficit has been creeping up every year regardless of economic conditions. New York City's liability status is the worst among all seven cities. Per capita short-term debt increased at a compound annual rate just under 25 percent and the adjustment in the debt/asset ratio is the worst of the seven cities.

The data shown in Tables 9 through 11 would seem to suggest weaknesses in the argument that a default by New York City would have a tremendous impact on the municipal bond market and that the ability of other cities to attract outside funds would be severely affected. Houston and Chicago issued almost no new short-term debt in the early 1970s, and Buffalo and Cleveland apparently tried to lower the amount of their debt outstanding or at least keep a stable borrowing rate. If the sample is representative, the data suggest that the debt need of other cities in the near future may well be small and certainly will not be as expansionary as that of New York City.

The data also show that Buffalo is very close to the crisis level. Its per capita expenditures are highest and tax-effort coefficient lowest among the seven cities considered. Likewise, Denver and Cleveland may also be approaching the critical stage. However, Denver's future may not be as gloomy as Buffalo's, considering its rapid growth in per capita income.

Turning now to New York City's own experience, we present in Tables 12 and 13 some demographic and financial data on what has contributed to the city's ill health. Table 12 shows the change in population and jobs in New York City during the period from 1970 through 1974. The decline in jobs began almost simultaneously with the decline in population. In 1970 the city population peaked at 7,895,000 and private sector employment at 3,182,000. By 1974 New York City's population had declined by 300,000 with a corresponding decline of 300,000 in jobs. The rate of decline in jobs was more than twice the rate of decline in population.

Table 11
ANNUAL COMPOUND GROWTH RATE IN FINANCIAL CHARACTERISTICS FOR SEVEN LARGE U.S. CITIES, 1970–73

	New York		Chicago		Detroit		Cleveland		Buffalo		Denver		Houston	
	Percent	Rank	Percent	Rank	Percent	Rank	Percent	Rank	Percent	Rank	Percent	Rank	Percent	Rank
Per capita total revenue	6.0	(6)	6.5	(4)	7.4	(2)	6.4	(5)	4.9	(7)	9.6	(1)	7.1	(3)
Per capita expenditure	14.0	(5)	10.0	(2)	13.9	(4)	7.5	(1)	15.5	(6)	16.8	(7)	10.9	(3)
Taxes/expenditures	−2.2	(4)	4.0	(1)	−3.9	(6)	−5.9	(7)	−2.7	(5)	−2.1	(3)	−2.0	(2)
Per capita short-term debt	24.7	(7)	0.4	(3)	5.1	(4)	10.2	(6)	−9.0	(1)	7.7	(5)	none	(2)
Bonded debt/assets	17.6	(7)	−10.9	(1)	−6.6	(2)	−3.6	(4)	11.1	(6)	−4.9	(3)	−3.0	(5)
Revenues/expenditures	0.3	(6)	5.0	(1)	2.7	(4)	3.3	(2)	1.7	(5)	−1.0	(7)	2.9	(3)
Sum of rankings		(35)		(12)		(22)		(25)		(30)		(26)		(18)

Source: Same as for Table 10. Calculations of rates of change are compounded annually and are based on rounded figures.

Table 12

POPULATION AND PRIVATE SECTOR EMPLOYMENT
IN NEW YORK CITY, 1960–74

| | Population | | | |
	Total (thousands)	65 years old and over (thousands)	Percent 65 years old and over	Private Jobs (thousands)
1960	7,782	817	10.5	3,130
1970	7,895	955	12.1	3,182
1971	7,886	967 a	12.3	3,040
1972	7,847	974 a	12.4	2,999
1973	7,664	964 a	12.6	2,964
1974	7,567	964 a	12.7	2,877

a Estimated.
Source: U.S. Congress, Congressional Budget Office, *New York City Fiscal Problem: Its Origins, Potential Repercussions, and Some Alternative Policy Responses,* October 10, 1975, p. 11.

The flight to the suburbs by middle and upper income groups further eroded the city tax base and thus the potential city revenues. The decline in the tax base did not, however, slow down the growth of city expenditures. Table 13 shows past growth rates of city expenditures, employment, earnings of city employees, and tax capacity.

It can be argued that the growth in city employment, at least in the past, was a by-product of the growth of population and thus of the growing demand for public services. Although New York City's population has been steadily declining, the pace of public employment increased in relation to employment in the private sector. In addition to the increase in public employment, there has been a rapid increase in the pay levels of public employees, both in relation to private sector earnings and in relation to changes in the consumer price index. The earnings of New York City's full-time employees rose 129 percent between 1961 and 1973, in contrast to an 85.5 percent increase in private sector earnings.

The average annual rate of growth of the earnings of municipal employees between 1961 and 1973 was 7.1 percent as compared to 5.3 percent in the private sector.[19] When the rate of increase in these

[19] *Intergovernmental Anti-Recession Assistance Act of 1975,* Hearings, pp. 157-184, passim.

Table 13

RATES OF GROWTH OF KEY FINANCIAL VARIABLES
IN NEW YORK CITY
(percent)

Financial Variable	Compound Annual Rate of Increase
Total expenditure (1965–76)	13.1
Tax receipts (1965–75)	6.8
Welfare and social service (1965–74)	19.3
Elementary and secondary education (1965–74)	11.2
Higher education (1965–74)	22.1
Hospital and health care (1965–74)	15.1
Public employment (1964–73)	4.3
Private employment (1965–74)	−0.3
Annual earning of city employees (1961–73)	7.1
Annual earning of private sector employees (1961–73)	5.3
Cost of living index (1965–75)	4.2
Nonlabor cost (1965–70)	
Police	7.7
Fire	7.7
Education	8.4
Labor cost (1965–70)	
Police	6.9
Fire	5.4
Education	9.5
Retirement program (actuarial basis) (1965–70)	
Police	6.6
Fire	6.2
Education	17.5

Source: Rate of growth of expenditure components, public employment and earnings are calculated for data given in *Intergovernmental Anti-Recession Assistance Act of 1975,* Hearings, pp. 157–184; total expenditure and receipts from New York City Comptroller's Report. For nonlabor costs, see D. Greytak and R. Dinkelmeyer, "The Components of Change in New York City's Non-Labor Costs—Fiscal Year 1965–70; Supplies, Materials, Equipment, and Contractual Services," Syracuse University, 1974, mimeo. For labor costs and retirement programs, see B. Jump, "Financing Public Employee Retirement Programs in New York City: Trends since 1965 and Projections to 1980," Syracuse University, 1975, mimeo.

earnings is compared with the average annual rate of change in the cost of living in New York, it will be found that city employees have had a real gain in average annual earning power of 3 percent. Municipal employees also benefit from a generous program of fringe benefits. In 1961 the five actuarial pension systems required city contribution of $168.1 million; in 1974–75 total payments budgeted by the city reached $973 million, an increase of 579 percent. Because of the short-run time horizon of most politicians, there is a tendency to substitute liberal fringe benefits for pay raises, inasmuch as fringe benefits do not give rise to immediate cash flow problems. New York City officials are no exception to this tendency. Labor settlements in New York City have been characterized by fringe benefits above those found in most other cities. Here again, New York City has been mortgaging its future. In 1972 the unfunded actuarial liabilities of the city's five pension systems were reported to be some $4.5 billion over the next sixty-five years.

Although the average earnings of city employees exceed average earnings in the private sector, city service has been criticized as inefficient and inadequate and the performance of its employees ranked low compared to private-sector performance. It has also been pointed out that despite an enormous increase in the number of police personnel in the last twenty-five years, there are no more—and probably fewer—patrolmen on the street today than there were in 1950. The rise in pay levels, especially when fringe benefits are included, is attributed by many to the power of labor unions. However, the city cannot be absolved from guilt in its submission to union demands, given the inadequacy of its tax revenue, the levels of private earnings, and the obvious fact that whatever commitment the city makes must sooner or later come out of tax revenues.

The data in Table 13 reaffirm the view that the New York City government was too generous in its treatment of its employees and of most of its citizens in the face of its incapacity to raise additional revenues. That is, given its revenue constraint (current and potential), the city in effect was granting pay increases, providing a level of services, and promising fringe benefits that required mortgaging the future. Unfortunately, the day of reckoning came sooner than had been expected (if indeed anyone ever expected it to come) when a skittish capital market refused to accommodate new city borrowing to cover budget deficiencies.

If inefficient public services, high labor costs, and union power contributed to the acceleration of city expenditures, public policy toward social services also contributed. A brief outline of the parts

Table 14
EXPLAINING PER CAPITA EXPENDITURES IN NEW YORK CITY
AND SIX OTHER U.S. CITIES, 1952–73

City	Constant	X_1	X_2	X_4
New York	−.2176	−.0001 a (−2.9870)	−.00001 (−.6663)	−.0893 a (3.6193)
Chicago	−.7315	.0003 a (1.8845)	.00002 a (1.9622)	.0005 (1.2978)
Detroit	.6014	.0008 a (14.1550)	−.00004 a (−5.1393)	−.0037 a (−1.7847)
Cleveland	.5635	.0001 (.1289)	−.00004 a (−3.3221)	.0033 a (2.0248)
Buffalo	.2353	.0026 a (5.0691)	−.00001 (−.4986)	.0051 b (1.6723)
Denver	.0559	.0021 a (3.0188)	−.00001 (−.3499)	.0704 a (3.5959)
Houston	.0457	.0011 a (1.9313)	−.00001 a (−2.9086)	.0025 a (7.0276)

a Significant at 5 percent level.　　　b Significant at 10 percent level.
Note: Values in the parentheses are student t statistics. See text for explanation.

of this public policy would include (1) the city university's tuition-free and open-admission policy for high school graduates, which now costs over $612 million a year, up from $42 million fifteen years ago; (2) subsidized housing for low- and middle-income tenants; (3) rent control policies which have led to the deterioration and subsequent abandonment of some 30,000 rental properties, with delinquent property taxes amounting to $220 million in fiscal year 1975; (4) costly transportation subsidies and the low-fare policy; and (5) expanded health facilities beyond city needs, with $800 in health cost per person on welfare and a hospital vacancy rate of 25 percent.

The Fiscal Behavior of Seven Major Cities. To gain further insight into the fiscal performance of New York City, it may be helpful at this point to look at key expenditures, tax efforts, and debt policies of New York City, along with those of six other major U.S. cities. Using time series data for the period 1952–72,[20] we have attempted

[20] The results obtained should be interpreted with care because of data deficiencies. The data sources are U.S. Department of Commerce, *Local Government in Metropolitan Areas*, 1950, 1960, 1970, Sales Management Inc., *Survey of Buying Power*, 1950-1973, and U.S. Department of Commerce, *Finances of Municipalities and Township Governments*, 1962-1972.

X_5	X_6	X_7	X_8	X_9	R^2
.9223 [a]	−.2114 [b]	.9378 [a]	10.7070 [a]	−.0001	.999
(5.8913)	(−1.4434)	(5.9900)	(2.2926)	(−.1936)	
.1392 [b]	−.0117	−.1651	.0354 [a]	.0007 [a]	.990
(1.7404)	(−.2250)	(−.2102)	(2.7762)	(2.1744)	
.1013 [b]	−.0263	.2916	.0043 [a]	−.0003 [b]	.996
(1.3905)	(−.7401)	(1.3216)	(2.2989)	(−1.3586)	
.1228	−.0748	.3692 [a]	.0014	−.6219 [b]	.983
(.8018)	(−.6326)	(2.1458)	(.5077)	(−1.5591)	
3.5892 [a]	.0801	−.4562 [a]	.0095 [a]	.0001	.995
(2.9209)	(.9594)	(−1.9475)	(2.0396)	(.2278)	
.0467	−.1811 [b]	−1.5025	.0040	−.0002	.989
(.0342)	(−1.3721)	(−.8746)	(.8205)	(−.3911)	
.0244	−.0254	.0000	.0033	.0001	.983
(.8446)	(−1.0933)	(0.0000)	(1.0796)	(.3820)	

to identify those factors that have contributed most to the change in budget levels in the different cities. A multiple regression technique was used to explain per capita expenditures as a dependent variable (X_3), using the following explanatory variables:

X_1 = federal and state aid per capita
X_2 = population per square mile
X_4 = per capita income
X_5 = capital expenditure/total expenditures
X_6 = own revenues/total expenditures
X_7 = per capita short-term bonds
X_8 = city employee/1,000 population
X_9 = general economic condition

The regression results are summarized in Table 14. Coefficients and student t-values (in the parentheses) are written below each variable. A glance at Table 14 reveals some significant findings. First, except for New York City, all six cities considered have been adjusting their per capita expenditures in line with changes in federal and state aid (X_1). This is shown by the positive signs of the coefficients for X_1. By contrast, New York City's expenditures have

27

kept on rising regardless of what happened to outside aid. Second, per capita income represented a significant explanatory variable. As per capita income rose so did per capita expenditures. This relationship was exhibited in all cities except New York and Detroit, where the coefficient of per capita income changes was found to be negative. In the case of Houston, for example, for every dollar increase in per capita income, per capita expenditures rose by about three cents. Third, the coefficient for per capita short-term debt (X_7) shows how much New York City's expenditures have depended upon debt financing. The New York City coefficient (as well as the t-value) was the highest among the seven cities.[21] Fourth, the ratio of city employees to population was also a significant factor in the growth of municipal expenditures. In all cities the coefficient for X_8 was positive with the highest value of the coefficient being that for New York City. Fifth, the value of the coefficient for the ratio of capital expenditures to total expenditures (X_5) in New York City and Buffalo is significant. Finally, the demographic variable, X_2 (population per square mile), played an interesting role in explaining the regression relationship: when the population density fell, per capita expenditures were found to rise in all cities but Chicago. It would seem that the expenditure level was rigid downward in many cities despite the decline in the population of central metropolitan areas. Chicago was the only city in the sample that can be seen to have adjusted its per capita expenditures as the city population fell.

3. The Impact of Default

What would happen if New York City were to default? And what does a municipal government default mean? A municipal government default is quite different from a corporate default, both in the attendant legal proceedings and in its scope. In a corporate default, all the assets held by the corporation must be sold at the time of default to satisfy the claims of creditors, and subsequently the corporation ceases to be a continuing entity. But a municipal government is not to be dissolved nor are its assets liquidated. Several municipalities did default during the 1930s—Detroit on February 14, 1933, and Grand Rapids in March 1933—yet these municipalities still exist.[22] When a municipality defaults the procedure set by

---

[21] It is interesting to note that Buffalo was able to increase per capita expenditures at the same time it was significantly reducing its short-term debt. The sign of the coefficient is negative. This also confirms the results shown in Table 10.
[22] On default by other cities, see Advisory Commission on Intergovernmental Relations, *City Financial Emergencies*, July 1973.

federal bankruptcy law should be followed, but the existing law is said to be inadequate in its provisions to deal with municipal defaults or to safeguard against the interruption of city services as creditors attempt to stake their claims on the municipal government revenues. President Ford, in his speech on October 29, urged Congress to modify the federal bankruptcy laws to give the federal courts "sufficient authority to preside over an orderly reorganization of New York City's financial affairs." [23] The proposal outlined by the President would (1) prevent New York City funds from being tied up by lawsuits, (2) provide the conditions for an orderly plan to be developed for payments to New York creditors over the long run, and (3) provide a way for new borrowing to be secured by pledging future revenues.[24] The President also said that in the event of default the federal government would work with the court to ensure that police, fire and other essential services are maintained.

Clearly the President's proposal does not solve the city's financial crisis. The intent and focus of the proposal is on providing the city with an orderly transition towards fiscal reform without the chaos that might accompany bankruptcy. To assess the impact of default on various sectors of the economy, as well as on New York City itself, is obviously a complex proceeding. Without full knowledge of the details of the new bankruptcy law, of the extent to which the federal government would in fact work with the court to minimize the unnecessary disorder resulting from the default, and of the importance of the psychological effect of New York City default on investor attitudes, our analysis must to some extent be deficient. Rather than speculate on these points, we will confine ourselves to the investigation of some of the financial aspects of default, using necessary but heroic assumptions about market behavior and the response to city default.

The Municipal Bond Market. A look at the bond market during October 1975 (at the height of the default crisis) seems to suggest that

---

[23] Text of President Ford's speech on New York crisis, *Wall Street Journal*, October 30, 1975.

[24] The proposed new Chapter 16 of the Bankruptcy Act would allow New York City, with state approval, to file a petition with the federal district court in New York stating the city's inability to meet its debt obligations as they mature. The petition would be accompanied by a city proposal on ways to work out adjustment of its debts and a program for placing the fiscal affairs of the city on a sound basis.

To meet the short-term needs of the city the court would be empowered to authorize debt certificates, covering new loans to the city which would be serviced out of future revenues.

Table 15
NATIONWIDE AVERAGE RATES ON DEBT SECURITIES AND MUNICIPAL BONDS AND RATIO OF MUNICIPAL RATE TO Aaa CORPORATE RATE, OCTOBER 1975

	High-Grade Non-municipals[a]	Municipals[b]	Municipals/Aaa Corporates
October 10, 1975	7.39	7.48	.856
October 17, 1975	7.08	7.29	.836
October 24, 1975	6.89	7.17	.824
October 29, 1975	6.76[c]	7.13[c]	.812
October 31, 1975	6.69	7.36	.839

[a] Average rate for 90-day CDs, prime commercial paper, prime bank acceptances, and Aaa corporates.
[b] Index of twenty municipal bonds.
[c] Estimated by the authors.
Source: *U.S. Financial Data* (Federal Reserve Bank of St. Louis), October 29 and November 5, 1975.

the market as a whole has been in the process of adjusting to default since the beginning of the year. In the week of October 17, when the city was barely able to avert default, the average interest rate on bonds other than municipal bonds dropped by 31 basis points from the week of October 10, and it dropped another 46 basis points in the following two weeks, reflecting the shift away from municipal to other "less risky" securities. Table 15 shows market adjustments during the month of October 1975.

In the two days following October 29, 1975, when President Ford reaffirmed his stand against a federal bailout for New York City and urged Congress to modify federal bankruptcy laws, the interest rate on bonds other than municipal bonds fell by seven basis points. Uncertainty about the city's fiscal viability apparently intensified the placing of funds in "riskless" Treasury and other "less risky" securities, driving the yields on these securities down. The ratio of the municipal bond rate to the Aaa corporate bond rate did fluctuate between October 10 and October 31, but the degree of variation was not especially abnormal: the variation in this ratio was around .36 during fiscal year 1974–75. We conclude that the shift was not into corporates.

It is instructive to examine the reaction of bond-market professionals when New York City so nearly defaulted on October 17,

1975.[25] From *Moody's Bond Survey*: "The possibility of a New York City note default last Friday led to brief unsettlement in tax-exempts. Other market sectors, however, showed little reaction." From the Chase Manhattan Bank *Money Market Report*: "On Friday, doubt over the timely payment of maturing New York City notes led to some price erosion, but there is still substantial improvement over the week." From *Business Week*: "New York barely escaped default today, but the municipal bond market held essentially firm. This casts some light on the favorite question of bond market analysts: To what extent do the current levels of municipal bond prices discount default?"

The trend of the municipal bond rate suggests the stability of the rate and also the stability of the ratio of the municipal bond rate to the Aaa corporate bond rate. Between 1960 and 1974, the ratio of yield on long-term municipal bonds to the yield on Aaa corporate bonds varied from a high of .852 (1964) to a low of .638 (1974) with annual variation of the rates of between .36 and .58. Over the year from October 1974 to October 1975, the municipal bond rate fluctuated between 6.5 percent and 7.7 percent and its standard deviation was .316. The ratio of the municipal rate to the rate of Aaa corporate bonds was between .690 and .856, with a standard deviation of .036.[26]

While yields in the municipal bond market as a whole have risen in relation to the yields in other markets, it would also appear that there is a trend toward greater selectivity within the market. Thus the spread between Aaa and Baa municipals in 1974 was 64 basis points (5.89 percent for Aaa, 6.53 percent for Baa), while for the first nine months of 1975 it was 113 basis points (6.36 percent for Aaa, 7.49 percent for Baa).[27] In the third quarter of 1975 the spread increased to 125 basis points (6.50 percent for Aaa, 7.75 percent for Baa).[28] It is particularly worth noting that the greatest increase in borrowing costs came in the Northeast. In particular, in the third quarter of 1975, the estimated increment in net interest costs (in basis points) as the result of credit erosion from the prospective New York City default was 45.5 for New York State, 54.8 for Pennsyl-

[25] These are taken from a larger number of comments (all tending in the same direction) quoted in "Statement of the Hon. William E. Simon . . . before the Subcommittee on Economic Stabilization . . . October 30, 1975" in Department of the Treasury, *News*, WS-440.

[26] *U.S. Financial Data* (Federal Reserve Bank of St. Louis), various issues.

[27] Ronald W. Forbes and John E. Petersen, *Costs of Credit Erosion in the Municipal Bond Market* (Washington, D. C.: Municipal Finance Officers Association, November 6, 1975), p. 9.

[28] Ibid., p. 11.

vania, New Jersey, and Puerto Rico, and 41.8 for New England. On the other hand, the increase was only 8.4 basis points for the North-Central region, and the highest in any region away from the East Coast was 21.4 in the Pacific area.[29]

From both the longer- and short-term trends, it is clear that yields on municipal issues have so far maintained their historic relationship to the yields on corporate issues. In 1974–75, yields on municipal securities edged up in relation to the Aaa corporate bond rate, but the trend began to change during October when the ratio of municipal rates to corporate rates fell close to its normal level, remaining within the normal range at month-end despite a temporary increase in response to the President's speech on October 29 (see Table 15). The municipal bond market has certainly increased its selectivity. Since some market adjustment has already taken place, the question is whether further adjustments will become necessary as the day of default approaches and how long the adjustment period may be expected to last. From trend analysis of the bond market it can be inferred that the length of the cyclical variation (short-term) of the interest rate is between four and six weeks.[30] Once default has occurred, uncertainties about default would be removed and a speedy adjustment in the bond market could take place. Those investors who are currently withholding funds from the bond market could be expected to reenter the market, thus improving liquidity, lowering borrowing costs, and raising the credibility of the municipal bond market.[31]

The Bondholders. What about the bondholders? To the extent that market adjustment takes place, holders of New York City's bonds will inevitably experience some capital loss. Because of the lack of data on the distribution of New York City bondholdings by type of holder or by income group, no distributional effects of these losses can be investigated here. Table 16 gives a rough estimate of the distribution of total municipal bondholdings by major holders.

As Table 16 shows, the banking sector as a whole holds $7.2 billion (or 60 percent) of the $12 billion New York City securities out-

---

[29] Ibid., pp. 18-19.

[30] Estimated by the spectral method used upon Fourier Transformation. For details, see K. Chu, *Principles of Econometrics*, 2nd ed. (Scranton, Pa.: Intext Educational Publishers, 1972); see also G. H. Moore and J. Shiskin, *Indicators of Business Expansion and Conditions* (New York: Columbia University Press for National Bureau of Economic Research, 1967).

[31] "Statement of the Hon. William E. Simon . . . before the Joint Economic Committee . . ., September 24, 1975" in Department of the Treasury, *News*, WS-386.

Table 16
HOLDERS OF STATE AND LOCAL BONDS, NATIONWIDE
($ billions)

Issuers	Twelve Large New York City Banks	All Other Commercial Banks	Individual Investors	Total Bonds Outstanding
New York City and Big Mac	2.0 [a]	5.2 [b]	4.8 [b]	12.0
New York State	1.5 [a]	n.a.	n.a.	16.5 [b]
Other states and municipalities	4.5	n.a.	n.a.	201.5
Total	8.0 [c]	100.4 [c]	122.0 [c]	230.0 [d]

Sources: [a] *Business Week*, October 20, 1975; [b] *Business Week*, September 22, 1975; [c] Federal Reserve Board; [d] *Los Angeles Times-Washington Post* News Service.

standing. Of this $7.2 billion, about $2 billion is held by twelve large New York City banks, and the rest, $5.2 billion, is held by all other commercial banks and financial institutions. The remaining New York City debt is believed to be held by individual investors.

Inspection of Table 16 reveals, first, that New York City bonds amount to only 5 percent of the total municipal bondholdings of commercial banks, and second, that large New York City banks hold less than 20 percent of the total outstanding New York City securities. A look at the financial position of these larger New York City banks seems to suggest that for six out of the twelve banks, the loss from default would not be catastrophic since the six have a low debt/equity ratio.[32] The other six New York City banks are likely to experience some difficulty. They are small and their total holdings of New York City securities are put at about 70 percent of their equity level. However, given the repeated statements by the chairman of the Board of Governors of the Federal Reserve System that the system will and can lend funds to member and nonmember banks whose solvency would be jeopardized by a city default, the data here seem to

[32] In New York City security holdings and debt/asset ratio, the figures are Bankers Trust, $410 million, 7.1 percent; Chase, $420 million, 9.9 percent; Chemical, $260 million, 10.9 percent; Citicorp, $280 million, 4.6 percent; Manufacturers, $160 million, 5.7 percent; and Morgan, $190 million, 7.3 percent. For the twelve large New York City banks, the debt/equity ratio is 25 percent and it is 35 percent for all other commercial banks. See *Newsletter* of Shields Roland, Inc., October 1, 1975 and U.S. Congress, Congressional Budget Office, *New York City's Fiscal Problem*, background paper no. 1, October 10, 1975.

suggest that a New York City default would not have any strong effect on the banking community. Furthermore, the Federal Deposit Insurance Corporation's contingency plan for lending funds to the banks affected rather than forcing them into bankruptcy should also help these institutions overcome the impact of default. With an appropriate liquidity policy put into effect, the danger of a sharp erosion of the banking sector's liquidity position would be substantially reduced.

It may be estimated that some 100,000 individuals hold $4.8 billion of New York City bonds.[33] Those individuals would experience some capital loss, to the extent that the market value of the defaulted bonds would fall. However, if we assume that the distribution of New York City bondholdings is similar to the distribution of total municipal bondholdings, then the majority of those holders will be concentrated in the high-income brackets where the value of tax exemption makes such holdings attractive. Because of the deductibility of capital losses against other income and the likelihood that such high-income individuals have other sources of income, the tax-loss-offset provision in the federal personal income tax system should reduce the impact of losses on these groups. Those investors in low-income brackets—especially former and older city employees—will lose and their losses cannot be minimized.

There is not much information available on what has happened to bondholders when municipal bonds have defaulted. From 1938 (the year following the passage of the Federal Municipal Bankruptcy Act of 1937) through 1966, the admitted debts of municipalities in bankruptcy came to just under $211 million, and the admitted losses to a little less than $77 million.[34] Of these admitted losses, approximately $70 million were on municipal bonds defaulting in the latter part of the Great Depression.[35] And most of these were settled either by deferred payment of principal and interest after bankruptcy or by the granting of an extension to prevent municipal bankruptcy.[36] The available data are not sufficient to permit the determination of realized yields on the defaulted bonds. But the experience with defaulted corporate bonds may be indicative, and here at least there are data available for the period from 1900 through 1943.

[33] Authors' calculations.

[34] George H. Hempel, *The Postwar Quality of State and Local Debt* (New York: Columbia University Press for the National Bureau of Economic Research, 1971), p. 25.

[35] Ibid., p. 24.

[36] Ibid., n. 13.

The realized yield on large corporate bond issues defaulting between 1900 and 1943 was 2.3 percent from first offering to final
settlement.[37] The expected yield at first offering for the average issue
that defaulted was 6.4 percent. The loss rate was therefore 4.1 percentage points (410 basis points). The lowest loss rates and the
highest realized yields occurred on Aaa corporates and on railroad
bonds. The realized yield on defaulted Aaa corporates was 3.1 percent, against an expected yield of 4.7 percent, for a loss rate of
1.6 percentage points. The realized yield on railroad bonds of all
ratings was 3.3 percent, against an expected yield of 6.1 percent, for
a loss rate of 2.8 percentage points. It may be noted that both these
groups had a positive realized yield from first offering to default,
the yield being 1.8 percent for Aaa corporates and 0.1 percent for
railroad bonds. These data suggest that the effect of default on the
holders of New York City obligations, though significant, especially
for those who depend on the bonds for income, would nevertheless
not be catastrophic.

The Economy. There are essentially two approaches one might take
in analyzing the effect of the New York City crisis on the economy.
One approach is to assume that New York City default is inevitable.
In this case, we may investigate the impact of default on the municipal
bond market, the spillover effect into other financial markets and into
the economy. Alternatively, we might assume that a New York City
default can be avoided through the adoption of one or more of the
following policies: (1) raising city taxes and cutting spending sufficiently to balance the budget; (2) restructuring of outstanding debt,
especially lengthening maturity of short-term debt; and (3) post-
ponement of the payment of maturing debt and interest, or accessi-
bility to the bond market for new short-term borrowing.

In the no-default case, the impact on the economy would be felt
through reduced city employment and purchases of goods and
services and perhaps through a higher municipal borrowing rate.
Although it is difficult to estimate the likely impact of New York
City default on the economy, nonetheless such an estimate is war-
ranted. The cost to society that would result from letting New York
City default must be guessed so that it can be compared to the cost
of alternative policies aimed at rescuing the city from default. Since
no precedent of default by a major city exists, and since we do not

[37] Harold G. Fraine, *The Valuation of Securities Holdings of Life Insurance Com-
panies* (Homewood, Ill.: Richard D. Irwin, 1962), Table 2-3, p. 35, from which
all data in this paragraph are taken.

have accurate techniques for predicting public responses to New York City default, our attempt at quantifying the impact must rest on excessively heroic assumptions. First, we will assume that default by New York City would neither precipitate a storm of bankruptcies in the private sector nor give rise to a wave of municipal defaults. This assumption may to some extent be justified on the grounds that the dire predictions about the future of the financial system following the bankruptcy of the Penn Central in 1970 did not materialize.[38] Second, we will assume that the municipal bond market and financial institutions holding New York City notes have already made considerable adjustment to the possibility of a city default. If we look at the holdings of municipal bonds by commercial banks, we will note a slowdown of bank purchases beginning in 1972 and culminating in total withdrawal from the municipal market during the first quarter of 1975.[39] The annual net change in commercial bank holdings of municipal bonds declined from $12.6 billion in 1972 to − $2.7 billion in the first quarter of 1975.[40] Between January and June 1975, municipal bond yields rose from 5.8 percent to 6.5 percent and commercial banks reentered the market with additional purchases of more than $6 billion.[41] Finally, we will assume that the Federal Reserve System would "assure the integrity of the nation's money supply." [42]

Given the above assumptions, and using the AEI Budget and Resource Allocation Projection model,[43] we present in this section two sets of results corresponding (1) to a New York default assumed

---

[38] It is worth noting that the anticipated fury of the "Penn Central credit crunch" was held off by the efforts of Felix Rohatyn, who is now engaged in trying to hold off the anticipated fury of the New York City credit crunch. See Charles D. Ellis, The Second Crash (New York: Simon and Schuster, 1973). Probably the greatest threat to the immediate stability of the financial system in the United States would come if a New York City default exacerbated an already difficult credit situation, but that would depend on the timing of the default.

[39] Unpublished flow of funds data from the Board of Governors of the Federal Reserve System (processed, August 19, 1975).

[40] However, with the unprecedented rise in the volume of municipal borrowing during January-June 1975 and also the rise in municipal bond yields during that period, commercial banks reentered the market with additional purchases of $6.9 billion during the second quarter of 1975.

[41] U.S. Financial Data, week ending October 15, 1975 (Federal Reserve Bank of St. Louis), p. 6.

[42] Paul W. McCracken, "New York City and the Economy," Wall Street Journal, October 17, 1975.

[43] The model is discussed in Paul N. Courant, William H. Branson, Attiat F. Ott and Roy Wyscarver, AEI's Budget and Resource Allocation Projection Model (Washington, D. C.: American Enterprise Institute, 1973).

to take place sometime between December 1975 and June 1976 and (2) to assumed fiscal action, including budget-cutting of $800 million a year for three years, and other arrangements necessary to avert default. In the two cases, the impact on the economy is measured as the difference from a base solution not incorporating default. The model simulation in the default case assumes that the ratio of the municipal bond rate to the Aaa corporate bond rate rises (over the base-solution value) by 1 percent a year during the period between 1976 and 1978.[44] Since the municipal bond rate enters into the construction equations and the demand for financial assets equations of the state-local government sector of the AEI model, a high municipal bond rate is expected to affect state and local borrowing for construction adversely, and thus to have an adverse effect on total purchases as well as on the sector's demand for financial assets. The spillover effects from the state-local government sector to the rest of the economy is expected to show up as affecting the aggregate level of demand for output and total state-local employment.

In the no-default case, we assume that state and local purchases of goods and services are cut back by $0.8 billion annually from fiscal 1976 through fiscal 1978 and that the municipal bond rate rises, but by only one-half the rise in the previous case—.05 percent. The spillover from the state/local sector component of the AEI model to aggregate demand and employment comes from three sources: (1) the reduction in state and local purchases—a component of aggregate demand; (2) the effect of state and local purchases on state and local employment; and (3) the effect of the municipal bond rate on construction outlays.

In Table 17 the result of the simulation is given. Also shown in the table are the results of two other macroeconomic impact simulations.[45] As the table indicates, the overall economic effects of default would be minimal, so long as our assumptions hold true. The difference in construction outlays between the default and no-default cases would be slightly more than $1 billion throughout the economy. That is to say, if the default of New York City did not set off a string of bankruptcies in a "credit crunch," it would reduce GNP by less than one-sixth of one percent.

---

[44] On this point, see Gerard Adams and James N. Savitt, "Macroeconomic Impact of New York City Default," prepared statement, Hearings of the House Budget Committee, October 23, 1975.

[45] For detail, see ibid., and Otto Eckstein, "Some Economic Implications of New York City Default," prepared statement, Hearings of the House Budget Committee, October 23, 1975.

Table 17
PROJECTED CHANGE IN TOTAL STATE AND LOCAL PURCHASES AND GNP, 1976–78, WITH AND WITHOUT NEW YORK CITY DEFAULT
(in $ billions and in percent)

Alternative Cases	1976	1977	1978
Case A: New York City Default (AEI-BRAP)			
Construction outlays for education [a]	$ − 1.121	$ − 1.198	$ − 1.262
Construction outlays for all others (excluding highways)	− 2.585	− 2.763	− 2.913
TOTAL	$ − 3.706	$ − 3.961	$ − 4.175
Reduction in state and local purchases as a percentage of total base-solution purchases	− 1.2	− 1.4	− 1.3
Case A: New York City Default (DRI) [b]			
Reduction in state and local government real outlays as a percentage of total	− .2		
Case A: New York City Default (Wharton Model) [c]			
State and local purchases	$ − 3.4	$ − 6.0	
Case B: New York City Budget Cuts to Prevent Default (AEI-BRAP)			
Construction outlays for education	$ − .561	$ − .599	$ − .631
Construction outlays for all others (excluding highways)	− 1.293	− 1.381	− 1.456
Budget cut to reduce deficits	− .800	− 1.0	− 1.0
TOTAL	$ − 2.654	$ − 2.980	$ − 3.087

[a] State and local purchases deflators for education construction (*PIE*) and others (*PIO*) are:
PIE_{58}: 2.76 (1976), 2.95 (1977), 3.0 (1978)
PIO_{58}: 2.79 (1976), 2.98 (1977), 3.1 (1978).

[b] DRI Simulation is given in prepared statement of Otto Eckstein, "Some Economic Implications of New York City Default," Hearings of the House Budget Committee, October 23, 1975.

[c] Wharton Simulation is given in prepared statement of Gerard Adams and James N. Savitt, "Macroeconomic Impact of New York City Default," Hearings of the House Budget Committee, October 23, 1975.

Source: AEI-BRAP model. For details on state and local sector equations, see David J. Ott, Attiat F. Ott, James A. Maxwell and J. Richard Aronson, *State-Local Finances in the Last Half of the 1970s* (American Enterprise Institute, 1975), Chapter 2.

4. Policy Options

Several policy options have been suggested to deal with the New York City financial crisis. These choices fall into one of the following categories: (1) federal aid, a loan guarantees program to avert city default or provide post-default aid; (2) self help, with restructured city debt and the use of city pension-fund assets as collateral for new borrowing; and (3) default.

Federal Aid. Although federal aid to the city can take several forms—loan guarantees, direct aid, or reissuance of municipal debt—the loan guarantees option seems to have captured the interest of most congressional committees, the governor of New York State and New York City officials. Two versions of a loan guarantees program were approved by two congressional committees. On November 3, 1975, the House Banking Committee approved legislation to authorize $7 billion in loan guarantees to New York City before or after default. The bill would establish a five-member federal board chaired by the secretary of the treasury to guarantee up to $2 billion of short-term borrowing by the Municipal Assistance Corporation. These loans would have to be paid off by October 1, 1978. The board also could guarantee long-term securities—up to $5 billion maturing by 1989 and up to $3 billion maturing by 1999.

As a condition for such guarantees, the city would have to submit a plan for balancing its budget by the end of fiscal 1978, and the state would have to agree to raise taxes sufficiently to cover up to one-third of the city's projected deficit. In addition, holders of city and Big Mac debt would have to accept lower interest rates and longer maturities on the debt they now hold. The city's employee pension funds would be subjected to renegotiation to scale down their costs.[46]

Earlier, the Senate Commerce Committee approved a bill that would permit the federal board to guarantee up to $4 billion of Municipal Assistance Corporation obligations the first year but would phase out the guarantee by 1980. The shorter time-horizon guarantees in the Senate bill are intended to help the city avert default—that is, to solve its short-term cash problem—rather than to provide post-default aid. The only post-default aid offered by the Senate bill is a guarantee of $500 million of three-month obligations to help the city pay for essential services.

---

[46] House Speaker Carl Albert indicated that AFL-CIO lobbyists opposed the loan guarantees bill because of the requirement for renegotiation of New York's employee pension funds. *Wall Street Journal*, November 7, 1975.

Several arguments are made for and against federal aid. The main arguments for federal aid are summarized here. First, it is argued that New York City plays a "vital" role in the national economy and the world economy. Its financial collapse would bring catastrophe to other municipal governments, the nation and the world. Second, it is argued that New York's problems are to some extent the result of federal policy towards large urban centers saddled with minorities and the disadvantaged. Third, it is argued that New York City contributes more to the federal government (in personal and corporate income taxes) than it receives. Federal aid to New York City should not be viewed as a federal "hand-out," but rather as a means of correcting an inequitable federal-state-local fiscal relationship. Fourth, it is argued that aid to New York City should take precedence over federal assistance to foreign nations.

The first of these has been discussed in the previous section. The second is clearly debatable. One might argue that the influx of minorities and the disadvantaged into New York City is a by-product of "generous" benefit levels provided by New York City and not the result of federal neglect. As we saw earlier, the level of New York City services benefitting minorities and the poor exceeds the levels of these services in other cities.

The last two reasons represent common fallacies on the role of the federal government in the economy and in federal-state-local fiscal relations. The fact that New York City contributes more to the federal treasury than it receives in aid cannot be taken seriously as an argument for a federal bailout. If one were to carry the New York City analogy to the individual taxpayer, a case could be made for restructuring the federal tax/expenditure policy away from low- and middle-income groups in favor of higher-income groups. Studies on the incidence of federal benefits and federal taxation clearly show that individuals with incomes below $15,000 benefit disproportionately from federal expenditures while contributing disproportionately little in federal taxes. Those with incomes above $15,000 bear most of the burden of federal taxes but receive little in federal benefits. The principle that governs the relationship of the federal government to individuals must equally apply to state and local governments, if federal distributional goals are to be effectively implemented.

Comparing federal foreign aid to federal aid to New York City or to any other municipal unit also involves a misconception as to the role of the federal government in the economy. The federal government engages in various activities that yield benefits to specific individuals, to groups of individuals, and to society as a whole. In allocating the federal dollar among various programs, the government attempts to maximize society's welfare in line with the preferences of its citizens (or of the majority of its citizens). If a majority favors spending more on income maintenance and less on defense (or the other way around), the budget dollar would be shifted to satisfy these preferences. The fact that the federal government spends some $4 billion on foreign assistance should mean that the majority of our citizens view the expenditures as augmenting their welfare. The reallocation of a budget dollar from foreign aid, defense, social security, or the like, to New York City must reflect society's preferences for such a reallocation and should not be made because the federal dollar should inherently go to New Yorkers and not to foreigners. It is not certain how many taxpayers now believe that the United States ought to shift its priorities toward rescuing New York City.

The argument against federal aid to New York City is likewise summarized here. First, federal aid to New York City would set a precedent for federal intervention in the affairs of local governments. Second, federal aid to New York City would open the door for federal aid to all municipalities that are (or appear to be) in financial difficulties. Third, federal guarantees of city bonds, especially if the new bonds were taxable, would provide large capital gains to present bondholders. Fourth, if helped out of its financial difficulty, New York City would not have the chance it now has to alter its long tradition of budgetary gimmickry. Fifth, if federal aid were to be limited to New York City, then the federal government would be rewarding irresponsible behavior at the expense of prudent and efficient management.

Federal bailout does not appear likely to serve the best interests of the city. There is no guarantee that the city will balance its budget and submit to fiscal discipline. As the President put it, "if they [the New York City officials] can scare the whole country into providing that alternative now, why shouldn't they be confident they can scare us again into providing it three years from now?" [47]

The prospects for federal aid to New York City do not now (November 11) look very promising. Most analysts believe there is only a small chance that loan guarantees legislation will be passed.[48]

---

[47] Text of the President's speech on October 29, 1975.

[48] A Harris poll released on November 5 said that 60 percent of Americans would favor federal guarantees if the city balances its budget and such a plan does not cost the taxpayers any money.

And if it is, the President has said that he will veto it. The prospects of overriding his veto would appear to be slim.

Self-Help. One promising alternative to federal aid is the plan recently outlined by Felix Rohatyn. This plan, released by Municipal Assistance Corporation officials, will raise $6 billion of new money. It involves (1) lengthening to ten years the maturities of $1.1 billion of Municipal Assistance Corporation bonds held by banks and due in three to five years (assuming cooperation from bondholders); (2) outright purchase by the city's pension funds of $1.87 billion of New York City securities due in three years; (3) asking state and city retirement funds to hold $1.3 billion of Municipal Assistance Corporation bonds; (4) a bank roll over of $550 million of New York City notes previously agreed to; (5) "bridge" loans from commercial banks of $1.5 billion for three months; and (6) an exchange by individual holders of $1.6 billion of city notes for 9 percent Municipal Assistance Corporation bonds due in fifteen years.[49]

Governor Carey has pointed out that since May the city has undertaken several steps toward solving its financial troubles, including the imposition of $300 million in new taxes,[50] tens of thousands of firings, cuts in services, and increases in transit fares from thirty-five to fifty cents. With or without federal aid, these "self-help" steps are prerequisites for any solution to New York City's financial problems. Given the dimensions of the problem, this option seems to us to be sound on both economic and political grounds. The city and the state are not impoverished: city and state pension funds own over $14 billion of assets and their faith in their city (whether demonstrated by allowing the city to borrow against their funds or by investing some of their assets in city or Municipal Assistance Corporation bonds) is essential to encourage others to have similar faith in New York City's pledges for financial reform. The city could increase its tax effort without damaging its financial capacity, if a system of user charges were instituted, if efficient housing policy were to replace the existing policy, and if there were better fiscal management than there is now. As we saw earlier, the city tax effort is low in comparison to that of other major U.S. cities.

---

[49] *Wall Street Journal*, November 7, 1975.

[50] Mayor Beame indicated in a speech to the National Press Club that taxes have been increased this year by $350 million on real estate and $330 million on business (speech, November 5, 1975), reported in *New York Times*, November 6, 1975.

Default. While New York City's default may not be as imminent as various public pronouncements seem to make it, it is not unlikely. The financial and economic effects of default have been discussed earlier and it should suffice here to point out that a necessary condition for an "orderly" default is the passage of the bankruptcy proposal outlined by the President on October 29, 1975.

The three options discussed above must all be viewed as short-term solutions to New York City's financial ill health. The city's crisis is not solely a cash-flow crisis that can be cured by $4 billion or $7 billion of federal aid. The roots are deep and, of necessity, the cure will be bitter and long. Over the next decade the financial solvency of the city will still be threatened if past trends continue. The city must undergo major surgery (or a host of major surgeries) to get back on the road to solvency. It must overhaul most of its spending and debt policies. There must be a reallocation of responsibility between the state and the city for certain services (as higher education and welfare). Other sources of revenue must be found and efficient budget management instituted.

Although clearly these are long-term goals, one cannot lose sight of the short-term problems. Here the first step toward a long-term solvency may be taken—that is, over the next few months, the city must demonstrate its willingness to solve its own problems. Beyond assuring that essential services must be maintained, the city should not seek a federal bailout under whatever label it may be packaged. If the city, with help from the state, can restructure its debt and restore faith in its ability to implement fiscal reform, the short-run crisis will be averted and the trend toward insolvency will be reversed.

Conclusion

Although there would undoubtedly be some ill effects from default— both in the restriction of credit and in economic hardship from the cessation of interest payments—it would appear that the effects have been exaggerated in the popular press. When a corporation defaults, it is likely that the corporation, in one way or another, will cease to exist; but no one is arguing that New York City would cease to exist, though it will certainly have its wings clipped even if there is no default.

In part, the conditions that have led to the city's financial crisis are endemic in almost all large cities in the United States. In part, they are peculiar to New York City, or very nearly so. There is no question that cities in the United States are generally suffering from a decline in their ability to support themselves and their citizens in the style to which they have become accustomed. But only in New York City has the decline in the ability for city self-support been coupled with excessively rapid growth in public expenditures *and* a high degree of long-term borrowing to pay short-term bills. The New York City experience can be contrasted with that of Chicago, where a shrinking tax base has been accompanied by a firm hold on public expenditures.

Of course, the answer to the question "What is to be done to prevent default?" depends on the answer to the question "What will happen if there is default?" as well as on the answer to the question "What will work?" It would appear that New York City is capable of saving itself, though it is not clear that the will is there, even if the way is. Still, city pension funds could be used to stave off default between December 1, 1975, and June 30, 1976, as could state pension funds. But this would represent only a temporary expedient, and it would be throwing good money after bad if there were no changes made in New York City's financial policies. Moreover, it is not at all sure that a pension fund manager who agreed to buy city obligations, or even Big Mac obligations, or even to allow his funds to be pledged against these obligations, would be acting as a prudent man.

We have argued here that default, though not something to be welcomed, would not be a catastrophe. The prevention of default is not, therefore, something requiring heroic measures. We have argued that there is a reasonable chance for New York City to hold off the expected default for the next six months, during which time steps can be taken to balance the city budget without the financial gimmickry of the past. Since this can be done—or at least since there is a reasonable chance it can be done—there is no need to do more. To put the argument in its simplest form: if New York City, with the Municipal Assistance Corporation and the Emergency Financial Control Board, can save itself, and if the consequences of its failure, while significant, would not be catastrophic, then there is no need for the federal government to step in.